YOUR KNOWLEDGE HAS VALUE

AF141558

- We will publish your bachelor's and
  master's thesis, essays and papers

- Your own eBook and book -
  sold worldwide in all relevant shops

- Earn money with each sale

Upload your text at www.GRIN.com
and publish for free

Paulo Lyimo, Salim Shaaban

Assessment of Forest condition at SUA-Kitulangalo forest reserve in Tanzania Miombo Woodland

GRIN Publishing

Imprint:

Copyright © 2015 GRIN Verlag GmbH
Print and binding: Books on Demand GmbH, Norderstedt Germany
ISBN: 978-3-656-88535-1

This book at GRIN:

http://www.grin.com/en/e-book/288083/assessment-of-forest-condition-at-sua-kitulangalo-forest-reserve-in-tanzania

GRIN - Your knowledge has value

Since its foundation in 1998, GRIN has specialized in publishing academic texts by students, college teachers and other academics as e-book and printed book. The website www.grin.com is an ideal platform for presenting term papers, final papers, scientific essays, dissertations and specialist books.

Assessment of Forest condition at SUA-Kitulangalo forest reserve in Tanzania Miombo Woodland

[1]Paulo J. Lyimo* and [2]Salim Shaaban
[2]Forestry Development Trust, P.O.Box 2,
Iringa, Tanzania

Abstract.
There is high deforestation and forest degradation estimated to be 400,000ha per annum between 1990 and 2013 in Tanzania forests. This is due to the increase in wood demand for energy and for industrial raw materials. The study was conducted to assess forest condition of Sokoine University of Agriculture - Kitulangalo Forest Reserve (SUA KFR). A total of 52 nested circular plots with radii of 2, 5, 10, and 15 m were used to collect data such as diameter at breast height (DBH), height and number of regenerants. Parameters computed for data analsysis were stem per ha (N), number of regenerants, dominant tree species, exotic species, basal area per ha (G) and volume (standing and removed) per ha (V). A total of 71 tree species were identified and recorded during inventory, the average 995±256 stems/ha of different tree species of various DBH class, the average basal area per ha, G=7.961268±0.8 m^2/ha and the mean standing total volume, 54.72914±11.3 m^3/ha were also computed. The mean total volume removed is 0.58 m^3 per ha which is very small compared to stocking. The dominant tree (richness) species includes *Julbernardia globiflora (17%)*, *Brachystegia speciformis (8%)*, *Acacia nigrescens (7%)*, *Acacia robusta (6%)*, *Albizzia harveyi (5%)*, *Scierocaryabirrea spp/ Caffra (5%)* and others i.e total of all *<5% (52%)*. The conducted inventory shown that there is an average of 6121±2777regenerants per ha. Only *Senna siamea* was identified as exotic species which planted beside the forest boundary as a demarcation. In general, forest structure parameters indicate that SUA KFR is in a good condition as have high species richness and stem density per ha. We recommend more research on assessing forest condition in other forests in order to be aware with their status and look forward for the proper management measures.

Keywords: forest, condition, miombo, stem per ha, basal area per ha, volume

Introduction

Miombo woodlands cover a vast area of south and central Africa. Stretching from the northernmost tip of South Africa up to Tanzania, and from Mozambique in the east to Angola in the west, they cover approximately 3.2 million km^2 (Scholes & Biggs 2004). Fires are the characteristic feature of the miombo woodlands. More detailed descriptions of miombo and how it is differentiated from other savanna or forest types are provided by Huntley (1982), Chidumayo (1993) and Frost (1996). Tanzania has about 48 million hectares of forests and woodlands which is about 55% of the total country land area (Mgoo, 2013). Out of this total area, almost two thirds consists of woodlands on public lands which lack proper management. Public lands are under enormous pressure from expansion of agricultural activities, livestock grazing, fires and other human activities. About 13 million hectares of this total forest area have been gazetted as forest reserves. Over 80 000 hectares of the

1

gazetted area is under plantation forestry and about 1.6 million hectares are under water catchment management. The forests offer habitat for wildlife, beekeeping, unique natural ecosystems and genetic resources. They are also an important economic base for the country's development. Miombo is characterized by *Brachystegia* and an understory of grasses, often growing on nutrient-poor soils derived from acid crystalline bedrock. In addition to *Brachystegia*, other important tree genera are *Julbernardia* and *Isoberlinia* (Campbell, Frost & Byron, 1996). Closely associated species of Miombo woodland are *Pterocorpus angolensis*, *Vangueria spp*, *Vitex spp*, *Combretum spp* and *Dalbergia spp*. The undergrowth is dominated by a heliophilous grass layer and forbs. Dominant grass species include *Hyparrhenia spp,Themeda triandra* and *Panicum maximum* and most common forbs are *Indigofera spp*. Patches of semievergreen forest (Kielland-Lund 1990) are more conspicuous in the western part of the reserve along the base of Kitulanghalo Hill where the tree layer is dominated by *Manilkarasulcata* and *Scodophloeusfischeri*. Common smaller trees are *Cola clavata* and *Strychnoshenningsii*.This tree species grow in areas where the climate is characterized by mean annual temperatures and precipitation of 18.0-23.1 ° C and 710–1365 mm (Frost, 1996).Miombo savannahs are home to important animal populations including elephants, lions, buffalos and antelopes and have high bird diversity (Campbell, Frost & Byron, 1996; Frost, 1996). Additionally, the miombo woodland ecosystem represents an important supply of fuel wood, fruits, poles and timber in villages, periurban and urban areas (Desanker et al., 1997; Sileshiet al., 2007).The Sokoine University of Agriculture - Kitulangalo forest reserve (SUA-KFR) covers 600ha out of the main Kitulangalo Forest Reserve comprised of 2638 ha. SUA-KFR is an important watershed management and soil conservation areas that contribute to urban water supply in Dar essalaam city through Sangasanga river which pours its water into Ngerengere river among the main tributaries of Ruvu river a major source of water to the city. Although the management of SUA KFR entails total conservation, there are noted areas for revenue collection like eco-tourism, beekeeping, collection and sell of non-wood products. The communities adjacent to SUA KFR in the two villages (Maseyu and Gwata) are collecting revenue from few researchers who visit the forest reserve. Biodiversity value, amelioration of climate, honey collection from tree boles, hunting of animals to increase food varieties, medicines, collections of vegetables, mushrooms, fruits and insects (sondo), fuelwood, construction materials and source of water for irrigation and drinking are potential values of SUA KFR. Eco-tourism provisions, local employment for casual labour and social values form important potentials to the local communities. Research and training together with the eco-tourism and carbon sequencing are potential for national and international communities. Being a natural forest of higher biodiversity values, SUA KFR has a great chance to qualify for Carbon credit. Despite of all these advantages offered by many forest still deforestation and forest degradation estimated to be 400,000ha per annum between 1990 and 2013 in Tanzania forests (FAO 2011; Mgoo 2013). There is need to be aware with the status of forests in order to take appropriate measures to ensure availability of forest resources for the benefit of present and future generations. This study was conducted to assess forest condition such as stem per ha, number of regenerants, dominant tree species, exotic species, basal area per ha, standing and removed volume per ha of SUA KFR.

2

Materials and methods

Study site

Kitulangalo is located at almost 35 km east of Morogoro municipality alongsideMorogoro–Dar es salaam high way, the major means of transport for the people and forest products to urban and commercial centers such as Dar es Salaam and Morogoro. Morogoro municipality is about 200km west of Dar es Salaam city.

The predominant feature in Kitulangalo is the Kitulangalo hill, which is about 762m above sea level situating at 06^0 39'-6^043'S and 37^057 -38^001'E on the left hand side when coming from Morogoro. The reserve is highly accessible throughout the year season since it's only few meters alongside the Morogoro-Dar es salaam highway, only 35km from the Morogoro municipality in Gwata-ujembe and Maseyu villages.This lead the forest to be open for easy access to the interior. (Figure 1)

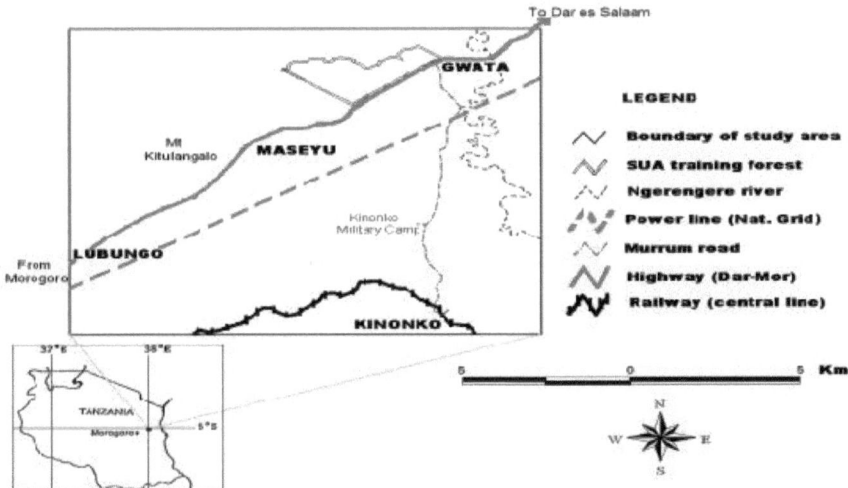

Figure 1: A map showing a location of SUA-KFR. (Source; Nduwamungu, J. et al., undated)

Data collection

The SUA – KFR map was used to guide the inventory operation in the field. Also systematic sampling design was used, where the plots were located in transects and distance between plots and between transect were 320m and 320m respectively.. The Concentric circular plots with various radiuses were adopted during inventory as shown figure 2 below;

3

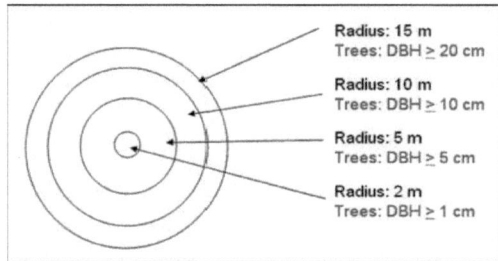

Radius: 15 m
Trees: DBH ≥ 20 cm

Radius: 10 m
Trees: DBH ≥ 10 cm

Radius: 5 m
Trees: DBH ≥ 5 cm

Radius: 2 m
Trees: DBH ≥ 1 cm

Figure 2: Concentric circular plots

Distance between transects and between plots

The longest forest boundary length is that along Dar es Salaam- Morogoro road, it is about 3200m and the whole SUA-KFR is to be laid out with 10 transects, hence the ground distance between transects is given by

3200m/ 10 = 320m

The first transect was laid at 160m from the boundary and all transects were laid at the forward bearing of 310^0 to the north in which its back bearing was 130^0

The number of plots in an inventory is usually calculated using data from a pilot study, the number of sampling plots is set to be 52 and plot area is set to be 0.070686 ha in which its radius is set to be 15m

The distance between plots is calculated as follows;

Let;

A = Area of the forest, 600ha

n = number of sample plots, 52 plots

Area (a) of the forest represented by plots, therefore is given by A/n, 10.3846ha ($103,846.15m^2$)

Distance between plot $=\sqrt{a}$, $\sqrt{103,846.15m^2} = 322m\sim320m$

Allocation of sample plots

Systematic allocation of the sample plots along transect line was applied. This was adapted due to the fact that among the advantages of systematic sampling design it ensures coverage of the whole population and it's easy in allocation of sampling plots. Plot shape chosen was circular plots since it serves time and increase accuracy. Measurementswere done into four concentric plots of radius 2m, 5m, 10m, 15m and.

The plot area in ha was calculated by $\pi r^2/10,000$

The result were that in radius of; 2m the area is $0.001257m^2$

5m is $0.00785m^2$

10m is $0.0314m^2$ and in

15m radius is $0.07069m^2$

Trees parameters measured and other forest resources

    i. In 2m radius we counted all seedlings, identified grasses and herbs.

    ii. In 5m radius we measured all trees DBH>1cm.

iii. In 10m radius we measured all trees with DBH$\geq$10cm.

iv. In 15m radius all trees and shrubs with DBH$\geq$20

v. All the stumps within the sample plot were measured.

The height of three trees, largest, medium and small in terms of DBH were measured using the hypsometer and their height were assumed respectively.

Data analysis

Data were analysed using Microsoft excel. Parameters computed were; Stem per ha (N), Basal area per ha (G) and Volume per ha (V). The number of stems per hectare were computed by using the following formula; N= $\sum$(i/A*n) where; 'N' = Stem density (stem count/ha) 'i' = Stem count per plot and 'A' = Plot area (ha) and 'n' = Number of plots. The basal Area (G) (m^2 per hectare) was calculated by using the following formula;

g = πDbh^2/4*10000

G=$\sum$(gi/A*n)

Where; Dbh = Diameter at breast height (cm); pi (π = 3.14159265).

A = Plot area (ha) 'n' = Number of plots and 'gi' = Basal area of a tree (m^2).

The Total volume per hectare was calculated using a formula, **V= $\sum$(0.000011972DBH$^{3.191672}$/plot area*n).**

The volume of stump was calculated using a formula;

$\sum$(V = 0.000011972DBH$^{3.191672}$/plot area*n).

The stump diameter was calculated first by a formula; Stump diameter=-1.003+ 0.87BD

Results and Discussion

Tree species composition at SUA-KFR

In SUA-KFR, there are 71 different tree species (Table 1) which have been observed and recorded in 52 plots which visited during field survey. The tree species was recorded by local and scientific name, where by the local names were provided by the indigenous and scientific names from botanist, literature review and other from internet search.

Table 1: Tree species at SUA KFR

No.	Local name	Botanical name
1	Kisasa	*Acacia goetzei*
2	Mkambala	*Acacia nigrescens*
3	mkandekande/kifunganyumbu	*Acacia nilotica*
4	Mkongowe	*Acacia robusta*
5	Mkongo	*Afzeliaquanzensis*
6	Mfuleta	*Albizziaanthelmintica*
7	Msisimisi	*Albizziaharveyi*

8	Mkenge/mkengepori/msagati	*Albizziapetersiana*
9	Mkingu/mnyanza	*Albizziaversicolor*
10	Mtomokwe/mtopetope	*Annonasenegalensis*
11	Msempelele	*Allophylusrubifolius*
12	Myombo	*Brachystegiaboehmii*
13	Msinzila/Chikundilekwima	*Brideliacathartica*
14	Mkundekunde	*Cassia abbreviata*
15	Mtutuma	*Catunaregumspinosa*
16	Mtulavula	*Clerodendrumglabrum*
17	Mlama-ng'ombe/mweupe	*Combretumadenogonium*
18	Mlama-goli	*Combretumcollinum*
19	Mlama-mweusi	*Combretummolle*
20	Mlama-mwekundu	*Combretumzeyheri*
21	Mtwinhi	*Commiphora africana*
22	Mkongolo	*Commiphoraeminii ssp. zimmermannii*
23	Mtwinhi	*Commiphorapteleifolia*
24	Mntindi	*Cussoniazimmermannii*
25	Mpingo/mhingo	*Dalbergiamelanoxylon*
26	Mzezegele	*Dalbergianitidula*
27	Kikulagembe	*Dichrostachyscinerea*
28	Msungura	*Diospyrossonsolotae*
29	Mkulwi	*Diospyroskirkii*
30	Mdaha	*Diospyros sp.*
31	Msofu	*Diospyroszombensis*
32	Mtogo	*Diplorhnchuscondylocarpon*

33	Msiga	*Doberaloranthifolia*
34	Msosowana/Mlwati	*Dombeyarotundifolia*
35	Mkwambe	*Flueggeavirosa*
36	Mkole	*Grewia bicolor*
37	Mkongodeka/mdandebande	*Grewiaectasicarpa*
38	Mbaazipori	*Indigofera spp.*
39	Mhnondolo	*Julbernardia globiflora*
40	Mhindi-pori	*Lanneaschimperi*
41	Mumbu	*Lanneaschweinfurthii*
42	Mfumbili	*Lonchocarpusbussei*
43	Mfumbili	*Lonchocarpuscapassa*
44	Mfumbili-mlima	*Millettiasacleuxii*
45	Mnenekanda	*Ochnamacrocalyx*
46	kilumbulumbu	*Ormacarpumkirkii*
47	Mgovu/mgomvu	*Pteleopsismyrtifolia*
48	Mninga/Mhagata	*Pteleopsisangolensis*
49	Msolo	*Pseudolachnostylismaprouneifolia*
50	Mng'ongo	*Sclerocaryabirrea ssp. Caffra*
51	Mhande	*Scrodophleousfischeri*
52	Mharaka/msalaka	*Spirostachys africana*
53	Mhembeti	*Sterculiaquinqueloba*
54	Mtonga	*Strychnosspinosa*
55	Mkwaju/Mkwezu	*Tamarindusindica*
56	Mtanga/Mfumba	*Terminaliasericea*
57	Mnyamafu/Mulyampofu	*Turraeanilotica*
58	Mzindanguruwe	*Uvaria sp.*

59	Mnyenye	*Xeroderrisstuhlmanii*
60	Mhingi/Mtwindi	*Ximeniacaffra*
61	Mnyangwe/mgagawe	*Ziziphusmucronata*
62	Mzeza	*Dalbergiaboehemii*
63	Kihale	*Haplocoeluminoploeum*
64	Mgama	*Manilkara discolor*
65	Msoto	*Dombeyarotundifolia*
66	Mngoji	*Pteleopsismyrtifolia*
67	Mjengaua	*Ekebergiacapensis*
68	Mgaluka	*Bosciasalicifolia*
69	King'angala	*Unknown*
70	Mtitu	*Caseariabattiscombei*
71	Mninga-maji	*Pterocarpusmildbraedii*

Number of stems per hectare (N)

The average number of stems per hectare was 995±256 stems/ha of different tree species of various DBH Class. These results are comparable with other studies in miombo woodland elsewhere in the country as shown in table 2 below. However the observed stems per hectare among the plots are different in terms of number, size and shape from that reported in other studies in Duru-Haitemba, Kitulangalo GFR and Urumwa FR (Nuru*et al* (2009; Malimbwi, 2003; Chamshama *et al.*, 2004; Zahabu, 2008). This difference is because sampling intensity (0.6%),, financial and human resources were not adequate for working with many plots more than 52 plots.

8

Table 2: Stand parameters from various studies in miombo woodland

Author	Forest name	N	G (M²/ha)	V (M³/ha)
Chamshama*et al* (2004)	Kitulangalo GFR	1085	9	76
Chamshama *et al* (2004)	KSUATFR	1027	8.95	76.02
Njana (2008)	Urumwa FR	583	8.54	58.41
Nuru*et al* (2009)	Urumwa FR	642	8.7	59.73
Malimbwi*et al* (2002)	Handen Hill	355	11.2	108.99
Malibwi (2003)	DuruHaitemba	1988	12.41	97.32
Zahabu (2008)	Kimunyu	701 - 845	7.9 - 8.8	72.4 - 88.2

The observed distribution of number of stems per hectare follows an inverse 'J'-shaped trend for each plot (figure 3), which is common for natural forests with active regeneration (Phillip, 1983) and recruitment. This trend happened due to previous disturbances from deforestation for charcoal making and wildfires that opened the woodland canopy and made way for more regeneration. Miombo woodland colonizes faster and more densely after disturbances because the woodland floor is exposed to sunlight and competition among woody plants is minimized (Campbell, 1996).Generally there is low number of larger sized trees in all plots. The observed low stocking of large trees is largely explained by the nature of the woodlands themselves, previous deforestation before the beginning of patrol in 2005 together with some illegal harvesting taking place in forest.

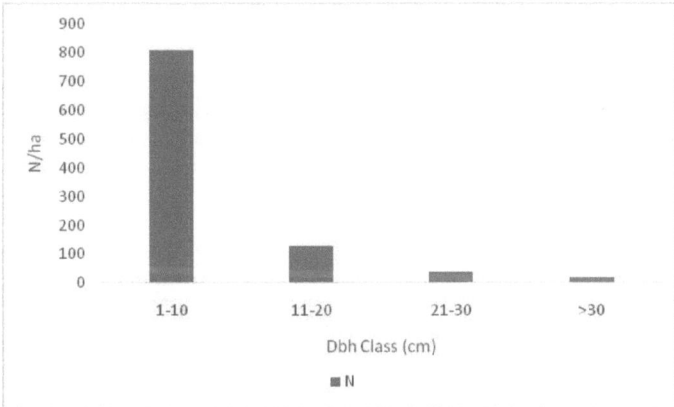

Figure 3; Distribution of number of stems per DBH class

Stand basal per ha (G)
The average basal areas per ha was **G=7.961268±0.8** m²/ha in SUA – KFR. This is contrary to 10.27 m²/ha of Zahabu, 2000.The result shows differences due to time constraints which lead to low accuracy. The forest appears to have two distinct canopy strata – the upper and a lower canopy. The two strata can easily be distinguished in the forest, which explains the bimodal distribution shown in Figure 4.

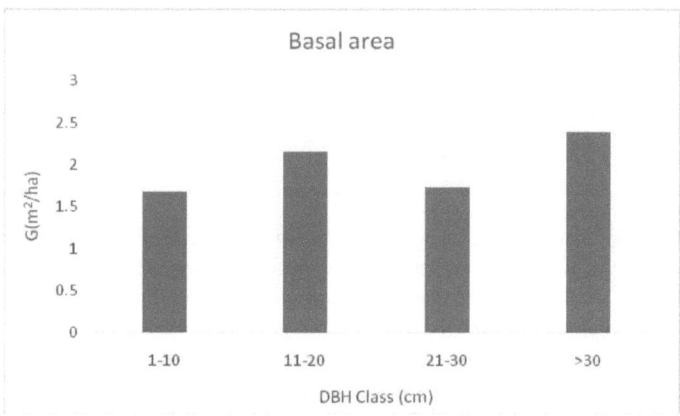

Figure 4 .frequency distribution of basal area in SUA-KFR

Total standing volume per hectare (V)
The mean total volume is 54.72914±11.3 m³/ha compared to 78.8 m³/ha by Dhahabu, 2000. Other reported standing volume from various studies in Tanzania are as shown in
Table 2.The difference could be caused by fragmentation of the forest due to new road construction, climatic stress and inaccuracy due to the fact that the instruments like compass were not enough resulting in poor alignment of the sample plots. The distribution of volume in SUAK FR show a J- shaped trend as expected in natural forest with good regeneration and recruitment. The distribution of volume in SUA KFR, show J shaped trend as expected in natural forest as shown in figure 5 below. .

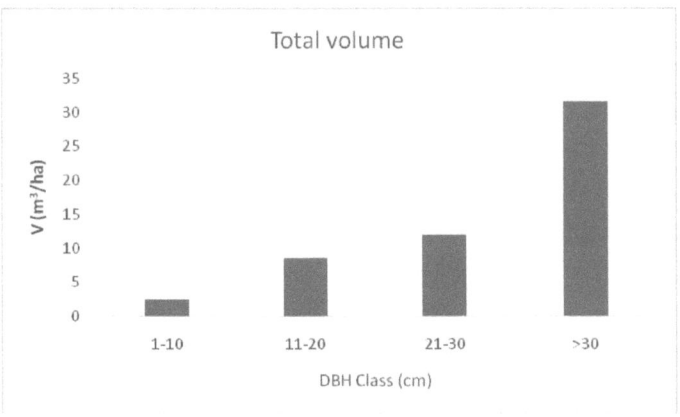

Figure 5: The distribution of volume/ hectare in SUA KFR

Volume for removed trees (V)

The total volumes per hectare of removed/harvested were determined so as to compare it with standing stocks using existing stumps. The total tree volume removed was 0.58 m^3 per ha which is very small compared to stocking. Also it was identified during field inventory at SUA-KFR that there were dead stumps some of which were due to fire, illegal timber harvesting, fuel wood and effect of grazing, but also some trees fall naturally due to lack of enough rainfall leading to poor survival Therefore the average percentages of trees volume removed and tree volume stock were 1% and 99% respectively.

The stump parameters observed in the SUA-KFR are as shown table 3 below

Table 3: Volume and Number of stumps removed

Species code	Scientific name	Local name	Sum of N	Sum of V
79	*Julbernardia globiflora*	Mhnondolo	3.54	0.27
110	*Spirostachys africana*	Mharaka/Msalaka	1.09	0.12
49	*Deispyroskirkii*	Mmoyomoyo	0.82	0.07
33	*Clerodendrumglabrum*	Mtulavula	0.54	0.03
2	*Adansoniadigitata*	Kisasa	2.18	0.03
45	*Croton sp.*	Mkambaku	0.27	0.02
21	*Brackenridgeazanguebaria*	?	0.27	0.02
120	*Terminaliasericea*	Mtanga/Mfumba	0.54	0.01
25	*Brachystegiaspeciformis*	Mbonha	0.82	0.00
4	*Acacia goetzeissp*	Msese	0.54	0.00
70	*Grewia bicolor*	Mkole	0.54	0.00
99	*Ozoroainsignis*	Mkomachuma	0.54	0.00
36	*Combretumpadoides*	Mkungalungo	0.27	0.00
104	*Pseudolachnostylismaproun*	Msolo	0.27	0.00

	eifolia			
56	*Diplorhnchuscondylocarpon*	Msiga	0.27	0.00
137	Unknown		0.27	0.00
Grand Total			12.79	0.58

Species dominance (richness and abundance) at SUA-KFR

The species composition dominance is the collection of all plant species that characterize the vegetation (Martin 1996). The most common measure of dominance composition is richness (the number of different species) and abundance (the number of individuals per species found in specified area). The composition of miombo woodlands appears to be relatively uniform over large regions suggesting a broad similarity in key environmental conditions (Frost 1996). Currently the dominant tree species includes *Julbernardia globiflora, Brachystegiaspeciformis, Acacia nigrescens, Acacia robusta, Scierocaryabirrea spp/ Caffra and Albizziaharveyi*. The abundance of dominant species is summarized in the pie chart below. These trees are used for timber production, medicinal values, and charcoal making, firewood and construction materials.

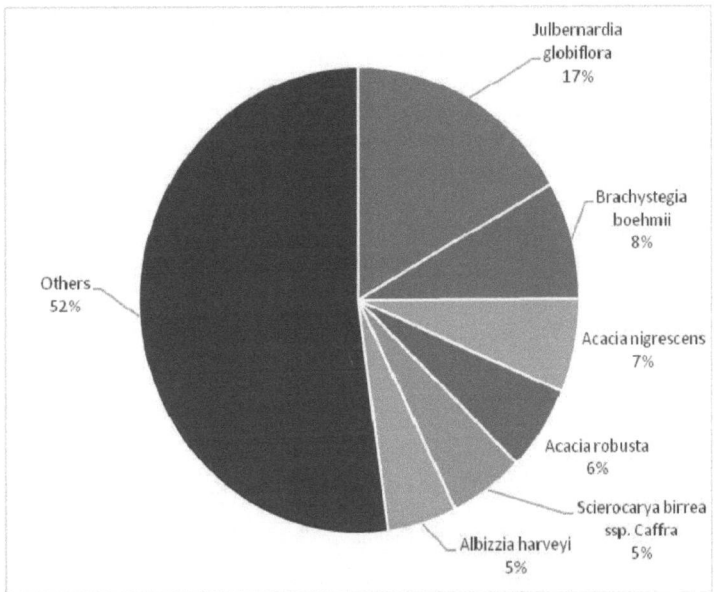

Figure 7.Species composition and dominance at SUA KFT 2013

Regeneration potential

The conducted inventory shows that there was 6121±2777regenerants per ha in SUA – KFR. The high number of regenerants in the public lands may be explained by disturbances on forest vegetation. Miombo species regenerate largely through coppice re-growth and root suckers rather than seeds, (Robertson, 1984 in Campbell, 1996). Chidumayo (1988) observed that stumps of miombo woodland have the ability to produce root suckers. Although seeds of majority of miombo trees and shrubs also germinate immediately after dispersal when there is enough moisture.tree density in re-growth of miombo woodland decreases with time due to moisture and heat stress. Shoot die-backs for cost miombo seedlings is common due to water stress and fire

Most regenerants die due to fire; this was investigated during inventory as a result very few saplings survive the miombo stresses. Fire had earlier mentioned by Kiellend-Lund, (1994) as the major cause of tree mortality in miombo woodlands. On other hand, fire was found to be the major ecological factors, which lead to development of miombo woodland (Lawton 1978). The impact of fire on miombo depends on time and frequency of burning and on the flammable biomass. Burning experiment in Ndola (Zambia) from 1933 Trapbell (1959) reported that repeated late and hot fires may destroy the woodland. About 50% of the trees died after 11 years of late fires. While early burning allowed maintained regeneration, complete protection leads to the development of a more closed partly evergreen forest.

Exotic species

Only *Senna siamea*was identified during inventory which planted beside the forest boundary as a demarcation but from ecological restoration point of view in the forest reserve it is advisable to use natural species (indigenous) to the area.

Table 2: Regenerants at SUA KFR

Local name	Botanical name	No. of seedlings
Kikulagembe	*Dichrostachyscinerea*	1882.3
Mhnondolo	*Julbernardia globiflora*	1117.1
Mlama-mweusi	*Combretummolle*	382.6
Mpingo/mhingo	*Dalbergiamelanoxylon*	290.8
Msempelele	*Allophylusrubifolius*	275.5
Mlama-ng'ombe/mweupe	*Combretumadenogonium*	214.2
Msinzila/Chikundilekwima	*Brideliacathartica*	183.6
Mkwambe	*Flueggeavirosa*	183.6

Msoto	*Dombeyarotundifolia*	168.3
Mfumbili	*Lonchocarpusbussei*	153.0
Mtanga/Mfumba	*Terminaliasericea*	153.0
Mdaha	*Diospyros sp.*	137.7
Myombo	*Brachystegiaboehmii*	122.4
Msosowana/Mlwati	*Dombeyarotundifolia*	91.8
Mhande	*Scrodophleousfischeri*	91.8
Mharaka/msalaka	*Spirostachys africana*	76.5
Kisasa	*Acacia goetzei*	45.9
Mgude/Mfune	*Sterculiaappendiculata*	45.9
Kihale	*Haplocoeluminoploeum*	45.9
Mkambala	*Acacia nigrescens*	30.6
Msisimisi	*Albizziaharveyi*	30.6
Mseni	*Brachystegiamicrophyla*	30.6
Mkole	*Grewia bicolor*	30.6
Mbaazipori	*Indigofera spp.*	30.6
Mninga/Mhagata	*Pteleopsisangolensis*	30.6
Mkwaju/Mkwezu	*Tamarindusindica*	30.6
Mhingi/Mtwindi	*Ximeniacaffra*	30.6
Mkilika	*Ehretiaamoena*	30.6
Mgulika	*Bosciasalicifolia*	15.3
Mkundekunde	*Cassia abbreviata*	15.3
Mtulavula	*Clerodendrumglabrum*	15.3
Mlama-mwekundu	*Combretumzeyheri*	15.3

Mtwinhi	*Commiphora africana*	15.3
Msungura	*Diospyrossonsolotae*	15.3
Mtogo	*Diplorhnchuscondylocarpon*	15.3
Mbaazipori	*Indigofera spp.*	15.3
Msolo	*Pseudolachnostylismaprouneifolia*	15.3
Mkusu	*Uapacakirkiana*	15.3
Mfurwe	?	15.3
Mtitu	*Caseariabattiscombei*	15.3
Mfumbili	*Lonchocarpusbussei*	6121.3

Conclusion and recommendation

This study reports findings of forest condition of SUA-KFR in terms of Stem per ha (N), Number of regenerants, dominant tree species, Basal area per ha (G) and Volume per ha (V). A total of 148 tree species were identified and recorded during inventory. Also there are 995±256 stems/ha of different tree species of various DBH Class, the average basal area per ha is G=7.961268±0.8 m²/ha, the mean standing total volume is 54.72914±11.3 m³/ha. The mean total volume removed is 0.58 m³ per ha which is very small compared to stocking. The dominant tree species includes *Julbernardia globiflora, Brachystegia speciformis, Acacia nigrescens, Acacia robusta, Scierocaryabirrea spp/ Caffra and Albizziaharveyi*. The result shows that there is an average of 6121±2777regenerants per ha. Only *Senna siamea* was identified as exotic species which planted beside the forest boundary as a demarcation. Generally, forest structure parameters indicate that the SUA-KFR is in a good condition due to high species richness, number of stem per ha, standing volume and basal area. We recommend more research on assessing forest condition such as species richness and diversity, number of stems per ha, basal area, volume, number of regenerants, exotic species and dominant tree species in other forests especially in miombo ecosystem due to its advantage to the community as indicated earlier in order to be aware with their status and look forward for the proper management measures.

Acknowledgements

We thank the Government of Tanzania (High Education Student Loan Board) through Faculty of Forestry and Nature Conservation, Sokoine University of Agriculture for the financial support. Special acknowdgement goes to BSc. Forestry class of 2010/2013 for their peacefull cooperation throughout this study. Lastly, we would like to thankfull to staff members of Department of Forest Mensuration and management Prof. G.C Kajembe, Dr. Eliakimu Zahabu, Dr. J.Z Katani, Dr. G. Mbeyale and Dr. Emmanuel Nzunda for their tireless and supportive idea throughout this work.

References:

Campbell, B. M. 1996. *The Miombo in Transition.Woodlands and Welfare in Africa.* Centre for International Forestry Research (CIFOR), Bogor, Indonesia, pp 11-57.

Campbell, B., Frost, P. & Byron, N. (1996) Miombo woodlands and their use: overview and key issues. In: The Miombo in Transition: Woodlands and Welfare in Africa (Ed. B. Campbell). CIFOR, Bogor.

Chamshama, S.A.O, Mugasha, A.G. and Zahabu, E., (2004). Biomass and volume estimation formiombo woodlands at Kitulangalo, Morogoro, Tanzania. *Southern African Forestry Journal* 200: 49-60.

Chidumayo, E.N. 1993. Responses of miombo to harvesting: ecology and management. SEI, Stockholm.132 pp.

Chidumayo, E. N., (1997). Miombo ecology and management.An introduction. Intermediate Technology Publications, London, UK.166 p.

Desanker, P.V., Frost, P.G.H., Justice, C.O. & Scholes, R.J. (Eds) (1997) Themiombo network framework for a terrestrial transect study of land-use and land-cover change in the miombo ecosystems of Central Africa. The International Geosphere-Biosphere Programme (IGBP) Report 41, Stockholm,Sweden.

FAO, 2007. Global Forest Resources Assessment 2010: Options and Recommendation for a Global Remote Sensing Survey of Forests. Forest Resources Assessment Programme working paper 141, FAO, Rome. 56p

Food and Agriculture Organization (FAO)., 2011. State of World's Forest. Rome

Frost, P. 1996. The ecology of miombo woodlands. In: Campbell, B. (ed.). The miombo in transition: woodlands and welfare in Africa. CIFOR, Bogor. pp. 11-58.

Gillah, P.K, Okitingati, F.B, Makonda, F.B.S, Kitojo, D.H and C.K Ruffo (2005). Increasing utilization and Market promotion of lesser known and lesser utilized Timber species in Tanzania. Publication number focal 3 – 2005

Huntley, B.J. 1982. Southern African savannas. In: Huntley, B.J. & Walker, B.H. (eds). Ecology of tropical savannas.Springer-Verlag, Heidelberg. pp. 101-119.

Kusaga M,M., (2010). Participatory forest carbon assessment in angai village land forest reserve in liwaledistrictLindi region, Tanzania ,MSc. Dissertation, SUA pp 167

Lawton, R. M., (1978). A study of the Dynamic Ecology of Zambia vegetation Journal of Ecology livelihoods of communities in Amani Nature Reserve, Muheza District, Tanzania. A   dissertation submitted in partial fulfilment of the requirements forthe degree of Masters of Science in the Management of Natural Resources for Sustainable Agriculture of the   Sokoine University of Agriculture Morogoro, Tanzania

Luoga, E.J, Witkowsiki, E.T.F and Balk, W.K (2000). Differential utilization and ethnobotany of tree in Kitulangalo Forest Reserve and surrounding Communal Lands.Eastern Tanzania Economic Botany.

Malimbwi R.E.,Zahabu E., Kajembe G. C., and Luoga E.J., (undated). Contribution of Charcoal Extraction to Deforestation: Experience from CHAPOSA Research Project.

Malimbwi, R.E.,(2007).Kitulangh'alo forest reserve: an overview.Themitmiombo project planning w/shop. 6-13th feb 2007.

Malimbwi, R.E., Solberg, B. and Luoga, E. (1994) Estimation of biomass and volume in miombo woodland of Kitulangalo Forest Reserve, Tanzania. Journal of TropicalForest Science 7 (2), 230 – 242.

Mbwambo L., Valkonen S., and KuuttiV., (2008). Structure and dynamics of miombo woodland stands at Kitulangalo Forest Reserve, Tanzania.

Ministry of Natural Resources and Tourism: (2002): Forest Act 1-48)

Ministry of Natural Resources and Tourism (2004) Forestry regulations

Ministry of Natural Resources and Tourism, (1998): National Forestry Policy:

Ministry of Natural Resources and Tourism (1999) Tourism Policy (PP 8-12)

Mgoo, S. J (2013). People in Forest Management: Experience from PFM in Tanzania. Presentation made at the Inter-Parliamentary Regional Hearing on Exemplary Forest Policies in Africa World Future Council held at Dar es Salaam, Tanzania from at 09-12 July 2013

Msanya, B. M.; Kimaro, D. N. and Shayo-Ngowi, A. J., (1995). Soils of Kitulangalo Forest Reserve Area. Morogoro District, Tanzania. SUA Department of Soil Science. 56 pp.

Munishi, P.K.T. & Shear, T.H. (2004) Carbon storage in afromontane rain forests of the eastern arc mountains of tanzania: their net contribution to atmospheric carbon. J. Trop. For. Sci. 16, 78–98.

Munishi, P.K.T., Mringi, S., Shirima, D.D. & Linda, S.K. (2010) The role of the miombo woodlands of the southern highlands of Tanzania as carbon sinks. J. Ecol. Nat. Environ. 2(12), 261–269.

Sileshi, G., Akinnifesi, F.K., Ajayi, O.C., Chakeredza, S., Kaonga, M. &Matakala, P.W. (2007) Contributions of agroforestry to ecosystem services in the miombo eco-region of Eastern and Southern Africa. J. Afr. Environ. Sci. Technol. 1, 068–080.

Scholes, R.J. & Biggs, R. 2004 (eds). Ecosystem services in southern Africa: a regional assessment. CSIR, Pretoria. 78 pp

URT (1998). Tanzania Forestry Policy, Forestry and Bee keeping. Division Ministry of National Resources and Tourism, Dar es Salaam – Tanzania

URT (2001).National Forest Programme in Tanzania. Ministry of National Resource and Tourism. Dar es Salaam – Tanzania.

Veltheim, T.andKijazi, M. 2002. Participatory Forest Management in the East Usambaras. Technical paper 61. Department of International Development Cooperation, Finland Metsahallitus – Forest and Park Service, MNRT/FBD

White, F. 1983. The vegetation of Africa. Natural Resources Research 20, UNESCO, Paris. 356 pp.

Zahabu, E (2000). Impact of Charcoal extraction to the Miombo woodlands. The case of Kitulangalo area Tanzania. MSc. Dissertation SUA, Unpublished

Zahabu, E. (2008) Sinks and sources: a strategy to involve forest communities in Tanzania in global climate policy. Dissertation submitted in University of Twente, Netherlands.